Words of Wisdom

A Collection of Apicultural Essays

WORDS OF WISDOM – A Collection of Apicultural Essays
by Ian Copinger.

ISBN: 978-1-908904-97-3

Northern Bee Books
Scout Bottom Farm
Mytholmroyd
Hebden Bridge
HX7 5JS (UK)

www.northernbeebooks.co.uk

Tel: +44 1422 882751

Design by Simon Paterson

Words of Wisdom

A Collection of Apicultural Essays

by

Ian Copinger

NB

NORTHERN BEE BOOKS

Scout Bottom Farm, Mytholmroyd, West Yorkshire

www.northernbeebooks.co.uk

Contents

I am pleased to give credit and thanks to my wife Patricia for the many improving suggestions which she made towards these stories without which some would not have appeared and others would have done less well.

Also for the cover sketch by my friend Tom Hume

Ian Copinger

.

For some years I have taken to writing an essay for the Essay Class in the National Honey Show. I was delighted to be receiving either a Third or a Very Highly Commended ticket. Then in 2009 The Art of Coarse Beekeeping won me first prize to be followed by another first for Bees and Darwin in 2010.

I would like to think that you will enjoy reading them as much as I enjoyed writing them. It certanly wasn't a case of dragging the words screaming from my soul.

Ian Copinger.

1
GM bees:
a concept for
the future

This one was written just for the fun of it during one of the peaks in the debate on GM crops.

A new queen arrives tomorrow from ABC, the Acme Bee Company. I was so disappointed with last years crop of hawthorn honey that I decided to change that queen. ABC are very good they can supply queens which will produce workers designed to gather in honey from one or two specific sources of flowers of your choice. I have one which does the early oil seed rape and the heather but ignores the late rape which overlaps the time that I am preparing to go to the moors and I don't want to risk contaminating the hive with rape honey.

I suppose we were all quite impressed by the slick salesman who came to one of our meetings to make an introductory presentation of the genetically modified honey bees, or to put it another way "sell". He was giving, he said, an illustrated presentation celebrating the conclusion of genetic experimentation which had resulted in the successful reconstruction of A. Mellifera to optionally specific ergometric patterns giving the customer a wider choice in the final composition of their end product, at which point three people walked out having decided that English was not this man's first language. Phrases flowed from his lips such as "increased readily harvestable assets" which we took to mean more honey and "the mono-sapidity of the product which would enhance the marketability to a new and wider consumer base"

ABC, he told us, had, as it were, taken the bee apart genetically and put it back together to a particular design which they could alter at will.

The bespoke honeybee had arrived and with it designer honey. The beekeeper pounced on the idea grasping the concept with both hands and swallowing the honeyed sales patter hook, line and queen excluder. ABC issued a list of high nectar producing flowers with their, now genetically controlled, flowering periods. From this list the beekeeper chose any combination he wished, handed over a cheque for an obscene number of "New Euro's", (speedily recoverable from the increased harvestable assets) and waited expectantly for the arrival of "your personally designed colony producer", (allow 28 days for delivery).

The need of beekeepers at different parts of the country had not been overlooked. Now that the South, more or less below that well known line from Birmingham to the Wash, had been designated a GM area while the North was "GM Organic", a concept far too difficult to explain in less than 50 pages, it was possible, indeed essential, to specify GM or GMO tendency for your queen. It had been decided that this was the best way to maintain the integrity of two areas. The alternative it seems would have required a 6 mile wide band of tarmac across the country.

I had also ordered one of the training colony queens since I did a lot of work taking the craft of beekeeping to schools in the area They were excellent because their workers had been so designed that they did not use their sting and so could be handled with total safety and without the need for expensive safety equipment for the whole class. It is a pity that one of the side effects of that modification has been that they don't collect much honey and so have to be fed almost continuously.

There was a panic when the Large Wax Moth, having been tempted further North by the warmer weather associated with the general climatic changes, reacted badly with the GMO pollen in the Ponte-fract licourice fields. The resulting flocks of mutants have now been driven into hiding in the depth of Sherwood Forest where once Robin Hood held sway and where local hunters now stalk them armed, of necessity, with shot guns.

On the good side I suppose you could say that Varroa as a parasitic enemy of the bee has now been largely wiped out. The drones, in whose cells they tended to breed, began eating the young before they themselves emerged from the cell. There are now so few of them about that BBKA were last heard of considering whether or not to ask the government to make them a protected species.

2
The bees at Beaurepaire

Written in 2001 largely at the request of Joe Graham the Editor of the American Bee Journal earning me $100.

The estate of Beaurepaire was created in the 13th century from land, about 1300 acres, given to the Prior and convent of Durham in the North East of England. The manor house on the estate was begun 1258 and became one of the most important houses in the County of Durham. It was extended over a period of probably 150 years to the typical 'E' plan of the mansion of the day. Its importance was such that Kings Edward I, II, and III visited it during their reigns.

Although it required a certain amount of rebuilding in 1346 after a visit from an invading Scots army of some 50,000 men on the eve of a locally held battle, the manor continued to flourish until 1540 when, with the dissolution of the monasteries, the office of Prior was lost and the use of the building changed. It suffered a considerable fire in the 17th century, probably the result of further visits from the invading Scots army in 1640 and 1644. By 1684 an attempt to obtain money to rebuild the house was made by Dean Granville who showed that a number of the courts, out houses and the walled garden were in ruins. By 1787 in was reported that "nothing but naked and distracted walls remain of this once beautiful place."

The walled gardens mentioned by Dean Granville lay to the east of the manor house. A 14th century inventory in the library of Durham Cathedral shows the number of bee colonies kept at Beaurepaire. Much further South in the country it was common to have deep alcoves called bee boles built into the thick walls of the garden to house skeps of bees. It was less common here and certainly there is no evidence of them in the 30 or so yards of the North wall which still remains. I suppose that the skeps, already well protected within the garden under the lea of the North wall, would be further protected by a thick sheaf of straw placed over them like a cloak.

The priory may have fallen into ruin and, except for a short stretch, the garden boundary gone and the garden now lost within a ploughed field but the bees are back.

The skeps have been replaced by British National hives and the British Black bee of years gone by has been replaced by a mongrel crossed with everything which was imported into the country after the Isle of Wight disease (Acarine) all but rid the country of bees around 1920. The apiary was re-established privately a few years ago and now usually 8 colonies benefit from the shelter of the remaining stretch of wall. With that to shelter them from the bitter North East wind and drifting snow the hives enjoy a warm Southern aspect and are almost guaranteed to come through the winter in reasonable strength. The whole of their foraging area is down to arable farming interspersed with old woodland and old meadow that the 14th century beekeepers would have recognised.

The bees wake up in the spring to coltsfoot on nearby reclaimed industrial land that was once Bearpark Colliery. Dandelions soon appear to give colour to roadside verges. Further West in Weardale dandelions grow in such profusion that a box of light delicately flavoured honey can be obtained from them. Here I have to fight the local council not to cut them all down two days after they come into flower.

The hard white honey that comes from the acres of oil seed rape is softened and tinged by the nectar and pollen which the bee add to it from the large old sycamores in the valley. Spring can be cold and wet so it is not unknown for the bees to eat most of that honey as fast as they bring it in. Hopefully field beans follow the rape and add a darker and stronger flavoured honey to the eventual harvest. The mixture we tend to call "Summer Flower Honey" is indeed a mixture depending on so many factors for it's unique flavour that will undoubtedly be different next year. Warm evenings might allow the flowering hedges of hawthorn to give off nectar. The bees might find those patches of white clover before the grass cutter does. There is still a sufficient acreage of old meadow with its flowering herbs to retain the flavour of honey that Rupert Brooke anticipated when he wrote, "Stands the Church clock at ten to three? And is there honey still for tea?"

Migratory beekeeping at Beaurepaire means putting half a dozen hives on the back of a Landrover and driving the twenty miles to the edge of the heather moors at Muggleswick. Traditionally this should be done on the 12th August. I really can't think why since The Glorious Twelfth signals the start of the grouse shooting season and I have no real desire to try to site hives on a moor with a dozen or so guns blasting grouse out of the sky around me. The thick, strong smelling heather honey is much prized by so many people and is sold at about double the price of summer flower honey.

By the time the bees are to be brought back from the heather the nights are drawing in and there is a chill in the evening air on the moor. Back in the home apiary it is time to take off the heather honey, or did they really just get enough in for their winter stores, put on the mouse guards, and treat for the inevitable Varroa.

I do not feed my bees ready for the winter but then I don't rob them of the honey which is theirs, I merely share a little of it with them. I shall not see my bees again until probably April but I am content that they are safe below their sheltering 13th century wall and anyway the Scots haven't invaded the place for years.

3
My biggest
beekeeping disaster
(2001)

This was my first entry to the NHS competition in 2001. I saw the schedule by chance and couldn't resist it. Every word is true although I have to admit I am not sure about Cyril. It got third prize.

"Disaster", noun, a sudden event bringing great damage, loss or destruction.

"Biggest", adjective, its all a matter of personal proportion really.

We were moving house. Hardly a disaster, actually it was something of a success but we lost quite a large garden and consequently I lost the apiary which was home to 5 nationals, one W.B.C. and a straw skep which had been running for three years. The skep rather falls into the category of small disaster since I threw a swarm into it with the intentions of removing them once they had drawn out some comb so that I could exhibit the skep on our County Show stand. The disaster of course was - now get them out.

So I needed a new apiary site. Not a problem, I was a local Police Sergeant, I had friends, I had influence, I had access to farms. I went to see one of these friends and explained my situation, his answer was as I expected, "Ian, help your self?" On a beautiful hot summers day I walked part of the diverse landscape of his farm. I ignored the ease of choosing a site too near the buildings for fear, would you believe, of a disaster should the stock or the farm hands get stung. I ignored, on advice, the South face of a paddock wall, "That's where we put the bull".

Beside one of the farm roads I came to a wide area of sycamore and chestnut on a South facing bank through which a band of some 30 feet had been cleared. It was fenced off with a high post and rail fence and sloped down a gentle stream. Looking down the short bank towards the stream I could feel the sun on my face and I knew that, without a further glance, I had found my new apiary site. Within two or three days a gateway had been formed in the fence, (I told you I had influence) and I began moving my bees onto the site. Meanwhile in the depths of Hades the demon of disaster chortled quietly to himself and waited, quietly and patiently for the result of his work to appear.

A month after the move the Bee's Officer, a good friend, chose to examine my bees. I proudly took him to my new site and we began the examination. Within moments it was apparent that all was not as it should be. My bees, once quiet and certainly of the non-following variety, were seriously annoyed. They flew at our faces, they banged into our veils and they committed suicide stinging our gloved hands. If they could have managed it by force of numbers they would have pushed us down into the stream. We retreated, back through the gate; they followed stinging as they came. We dived into our cars for shelter and watched as the bees reluctantly gave up their attempts to attack us any more. I couldn't understand it. Why should moving them have changed their tempers so radically? I looked up as though for guidance and for the first time saw the answer. Above me stretched seven electricity cables and the band of trees had been cut out to enable them to be slung huge pylons. My bees stood directly below the cables, well within the field of the current and they hated it and were showing that they hated it. They would have to be moved but not just yet because I really didn't have the time.

Meanwhile in the depths of Hades the demon of disaster, his name's Cyril by the way, sat back and chortled again. He hadn't finished yet.

Having settled my bees for the winter, with ample stores and mouse guards I determined to have a new site ready to move them to in the Spring. I had also moved my job and my daily journey now took me past the narrow valley half way along which stood my apiary. Every day I looked along it and thought of my bees. Winter came and temperatures dropped. No matter, my bees would stand the cold.

Quite early one cold winter's day I glanced along the valley to see that a light mist had settled there. If I gave it any thought at all it was that the weak winter sun would quickly move it. The mist was still there that evening. Next day the mist seemed to have thickened and looked even colder than it had on the previous day. By the end of the week it was quite clear that the light mist had become a freezing fog and had settled for the duration. One day I even fancied I could see a face in the swirls on the edge of the fog. Could it have been Cyril? The valley remained relentlessly filled with a freezing fog for three weeks as I drove past helpless in the face of nature!

Eventually the winter ended and I ventured out to look at the disaster that I already knew awaited me. The bees in all the hives were dead. My three-year-old skep colony, no honey but guaranteed early swarms, dead. At least it could now go on to the show stand and the comb lasted the test of small fingers poking at it for a few years. It and the other hives went home to be cleaned up in readiness for a resumption of hostilities.

"Disaster area", noun, an area officially declared to be the scene of a disaster and therefore qualified to receive emergency loans and supplies. I wish.

I went to another farm and found another site. Not an electricity pylon to be seen. It was a grassy knoll on a bank side fenced off no doubt against the dangers of stock falling over the twelve feet sheer face at the other side of the knoll. A quick adjustment of the fencing rails and I had a ready access. I was just banging a hive stand on the stony but grass-covered ground when my friend the farmer arrived. "The archaeologists have been out to look at that" he called up from the bottom of the twelve foot face. "They think it's an ice house"?. I walked round to join him and found my self looking at what, from this angle, was clearly an under ground building. I opened the huge iron doors that had been fitted across its front and coughed at the smell of the gasses rising from an open topped metal cess tank that filled the void.

Looking up I found myself gazing into the typical vaulted stone roof of a country mansion icehouse. The sun, filtered by the grass on top of the roof, shone through where some stones were missing, just about where, minutes earlier, I had been banging the roof with a hive stand. As I swung the doors closed I wondered what weight of beehive plus honey plus beekeeper might be required to send the whole package tumbling through the obviously unstable roof to an unsavoury death in a cess tank.

I thought of Cyril!!!!

4
I'll do it
tomorrow
(2002)

One of the judges remarked "We are often encouraged to start keeping bees but nobody suggests when we should stop!" I was quite happy at the time with my Very Highly Commended.

"Well I think it's about time to pack it in", the old man muttered to himself.

He was sitting on the window seat looking out over the long back lawn. The thin rays of the late winter sun filtered through the high and now largely leafless hedge and struck the side of the row of hives at the bottom of the garden. He had planted the hedge years ago to get the bees flying high, up and over the heads of his neighbours.

"Can it really be over 50 years?" he questioned his own memory. But, yes, he remembered the war he could not go to because of his job. He remembered the gift of bees and hives from the widow of the beekeeper who he had helped, from whom he had learned, and who had gone to war.

"Yes, they can go" he exclaimed to the empty room. He moved away to the big chair near the fire and beside his bookshelves lined with much thumbed copies of well-known authors. Here he had sat through snow filled winter evenings and re-read the words of Herrod-Hempsall and Manley.

The winter would be long without the first eager examination of the hives to look forward to. How much he had always enjoyed those first warm days of spring that allowed him entry to the hives. How relieved he was to find that a steadily expanding brood nest rewarded his autumn feeding and careful, if often late, preparation of the bees for winter.

He turned and picked one of the newer books from the shelf. The brightly coloured front of Ted Hooper's "Guide to Bees and Honey" brought back memories of summer days among the bees. The book lay unopened on his knee as his mind wandered over the fields to out apiaries near pleasant streams and heather moors with the cackle of grouse in the background.

The memory of the terrible winter of 47 almost made him shiver. The bees had to be fed all spring and well into the summer just to survive. They had been taken to the heather where, as if by a miracle, they had brought in such a quantity of rich deep coloured honey that he had kept a jar at the back of the cupboard all these years. He looked at it now and again and wished for another year like it but "please, without the winter".

The summer of 1956 when flowers had poured out nectar. When honey flowed, when colonies swarmed and when, because he had put off ringing his order in, he ran out of jars. Well there was bound to be a little hiccough now and again. There had been plenty of them of course when he had put off little jobs and as a result left himself with big problems to solve.

"Carpe diem" (Seize the day) his old gowned schoolmaster would cry out when he saw a chance being missed by some lethargic boy. Too often through his life "In crastinum differre" (To put off until tomorrow) had been the actual case.

He had always welcomed the opportunity to try different methods of swarm control combined with queen rearing, new little twists to a well trodden path. He recalled one attempt during which, because he had put off an essential step in the process, he had lost in one fell swoop, his breeder queen and two of her daughters. The loss also ruined his carefully drawn up heather honey production plan, a scheme that depended on having young queens laying by a given date in June. Having already put of the start of his queen rearing until the very last minute he was now unable to meet that deadline.

A coal shifted noisily on the fire and brought him back from his dreams. The fire was dying and it was time for bed. "If I get in touch with the secretary he can soon advertise them and some young bee-keeper can have the lot. I'll ring him tomorrow."

He stood up and walked towards the door. Passing the window he glanced out again as he had done for many a year. The moon reflecting off the whiteness of the hives now lit the garden.

"Yes, I'll do it, maybe tomorrow."

5
The beekeeper:
the bees' worst enemy
(2003)

At the 2003 N.H.S. I got a third for defending the beekeeper against the charge of being the bees' worst enemy.

A beekeeper is a person to whom friends call out, "Hello there. How are the bees?" Neither the beekeeper nor the friend are quite sure why the greeting is made in this manner and certainly if the beekeeper were to give a correct answer the enquirer would probably not understand a word of it. But how else is one to greet someone who keeps, apparently as pets of some sort, many tens, if not hundreds, of thousands of stinging insects.

"How's the little lady?" could lead to all sorts of misunderstandings especially when the reply is likely to be "I'm just about to give her the chop and get a younger one in" The friend may well be enquiring after the health and welfare of the beekeeper's wife but the beekeeper, single minded to the last, is thinking about the queen in his number 3 hive.

"Nice day isn't it?" hardly fits the bill. The beekeeper is rarely seen out of doors on anything other than a nice day. On wet, windy, cold days he skulks indoors where he can be seen fiddling with frames and wax in his shed or studying the television screen. He is not engrossed in some epic screen adventure he is reading weather forecasts and wondering when he can get at his bees again.

Just because he only ventures forth under the bright sun look not for signs of weathering or sunburn about his face. The beekeeper is forever clothed in all enveloping suits of gleaming white and a black veiled hat. And so he might be seen in the distance, bowed over an open hive in a cloud of smoke while he practises his art. Does this gentle description fit a person who is likely to be an enemy of the creature at the very centre of his craft?

'The craft', 'the art', both terms used profusely to describe the activity of beekeeping and, having become sufficiently expert, one might even be hailed as a 'bee master'. The wonder is that 17th century witch-hunts didn't connect the ability to manage bees, even in the most basic way, and to harvest honey from them with some unlawful arrangement with the devil. Then, by fire, terror and oppression, proceed to break any historic connection between man and bee leaving us all without the background relationship forged by early man.

Over the centuries man's relationship with bees has changed. The honey harvest of the hunter-gatherer was probably in a direct proportion to the number of stings he was able to stand. Only when civilization stabilized and the nomadic life gave way to settlements could the bee be managed within the formality of an apiary. Long before the pyramids were built beekeeping had become more formalized. Two thousands years ago Virgil wrote a description of an apiary designed for the benefit of the bee much of which still holds good today. However one is bound to wonder how many bullocks were actually suffocated to follow the Virgil method of the spontaneous generation of bees. In expanding their empire the Romans spread their knowledge and methods of management of the bee. Unfortunately they were starting from a base of incorrect information and conjecture based on country myth. For centuries this was re-published as fact by a succession of writers unable to properly observe the works of the bee enclosed within straw skeps.

This manner of housing the bee led to what now seems to be the most monstrous thing that man has done to the creature. The practice of killing colonies over a sulphur pit may well have been a matter of simple expediency. It saved the expense of feeding weakened colonies which might not survive the winter. Clearly bees must swarm at some time in order to increase the numbers of their race but the sulphur pit also destroyed all those colonies which, by their nature, did not swarm so often as others but rather tended to supersede their queen as necessary. In this way the beekeeper of years ago probably interfered with the Darwinian theory of evolution sufficiently to produce a bee bred to swarm and swarm again.

This unfortunate era slowed to an end with the introduction of the movable frame hive. Developed more for the benefit of the beekeeper than the bee it arrived in time for the bees' next serious problem, Isle of Wight disease or acarine. It doesn't seem too long ago that a very old lady described to me the "terrible day" when her brother announced that the bees were all dead. Acarine had found a weakness in the native British stock and wreaked havoc in apiaries throughout the country. No doubt a terrible time but hardly the fault of the beekeeper.

The beekeeper's reaction was natural and instant. Re-stock with bees from other countries which would be more resilient to the disease. However in doing so with such haste and so widely any chance of recovery for the native bee was ruined. Swamped with imports any surviving native colonies must have soon vanished, the subject of cross breeding into what is often today's mongrel.

That the bee is exploited to the full is not in dispute. It is clearly recognized as an exploited animal in an article in a 1984 Journal of the Institute of Biology. We harvest honey and gather the wax, a by-product of that harvest. We gather pollen in traps before it enters the hive. We falsely create a situation in which royal jelly is secreted into cells from which it can be harvested. Similarly, grids are inserted in hives from which propolis can be gathered. Even it's sting, or at least the venom from it, is used to treat rheumatism and to desensitize people who are allergic to bee stings. This hardly places us in the category of an 'enemy'.

The latest disaster to befall the bee is Varroa. In accounting for its spread we must each decide whether it has been by happenstance or enemy action. Bearing in mind the millions of years of its existence in, albeit parasitic but tolerant, harmony with its host, Apis Cerana, its recent spread across the world has been meteoric. Can the hand of man have been far away and isn't the hand most likely to have been that of a beekeeper moving infected colonies over vast distances to new areas.

The small hive beetle next threatens our bees. A creature which, in time, will undoubtedly land on our shores neatly packaged among some imported exotic fruit and ready to bulldoze its way through the nearest beehives. In the race to start a new disaster it will have to move quickly because it is clearly in competition with Apis mellifera capensis which is spreading it's way steadily through South Africa and the mite Tropilaelaps. The spread of these creatures will be speeded up not by the beekeeper but by the rapid movement of fruit and vegetable products around the world. Perhaps they could arrange to arrive in the same box of grapefruit. Now that would give us all something to think about.

Today the beekeeper is more than a harvester of honey. He is the protector of a species, a guardian of the environment. Great minds argue that without the bee the environment as a whole is at risk. Pollination would largely end, food plants die and, the food chain broken, the environment would collapse. Armed only with his knowledge of the craft and a few impregnated plastic strips the beekeeper tackles a parasitic mite which threatens the honeybee and apparently, by definition, life itself.

Meanwhile the chemical industry and, under some duress, the agricultural industry, in the never ending search for profit seeks to create a sterile environment in which cash producing crops can flourish. Arguing that herbicides do not kill bees they spray the fields and scatter destruction on any plant which might compete for the precious nitrogen, phosphate and potassium which their crops need. What then is left for pollinating insects to feed on? The various types of bumblebee are already much reduced and continue to be threatened. They are essential for pollination in temperatures when the honeybee cannot fly and for plants whose flower is so constructed as to require the size of the bumblebee to open it. Fields are sprayed to remove unnecessary flowers from the grass which will feed cattle leaving green deserts in which even the honeybee is hard pressed to find forage. Such massive loss of plant life can only serve to further diminish the number of bumblebees. The variation of flowers available has, in many places, so changed that the flavour of honey so loved by our forefathers has gone forever.

I will readily grant that, on a day-to-day basis, some of the actions of the beekeeper may be selfish and might not altogether be in the best interests of a particular colony. On the other hand would a beekeeper knowingly do harm to his own hobby which gives him pleasure and an income.

No, I think that for the true enemy of the honeybee you need to look elsewhere. Not necessarily very far, just a couple of paragraphs back might do. Having gained control of plants by chemicals and genetic modification how long will it be before the scientist seeks to mechanically manipulate the honeybee for their own ends? In fact I wonder if the reality of it is that they are already trying.

References

Virgil's Fourth Georgics and referred to in Tickner Edwardes 'The lore of the honey-bee', 1908

'Biologist', The Journal of the Institute of Biology. Volume 31. Number 4. September 1984.

6
The joys
of beekeeping
(2004)

The joys of beekeeping only got me a Very Highly Commended. A prize cards's a prize card. Don't knock it.

The winter is still with us but I need to think ahead and prepare for brighter days. Out with the catalogue and start writing an order for equipment. Wax, must replace some brood frames, order more super frames nearly ran out last year, might as well order my sections for the heather at the same time, add wax for them to the list, and replace two damaged queen excluders. There now, that's, HOW MUCH? Oh the joys of beekeeping.

Its Spring at last. The winter wind has turned Southwards, the snow has gone and the thermometer tells me I can now do more than stand in front of the hives watching the entrances and wondering. Finally my examination can be more than a cursory worrying peep under the covers.

I almost wish I hadn't bothered. No 2 hive is certainly queenless and 'how did that mouse get into No 4?'. At least there is no evidence of Varroa damage and I have recovered from worse over winter damage than that before. Oh the joys of beekeeping.

Kindhearted soul that I am I have offered to help a friend out with a bad tempered colony. It is 25o in the shade and I am drenched in sweat standing here in wellingtons, a full bee suit and household rubber gloves. I can hardly see because the inner curve of my glasses has filled up with the sweat which is running everywhere. We have been looking for the queen in one hive for fully half an hour and any minute now I am going to filter the whole lot through a queen excluder before lying down in a very inviting stream. Oh the joys of beekeeping.

I did eventually find the queen and the hive was requeened. My friend has just rung to tell me that the hive is running happily with the new queen, the bees are now calm, gentle and he has thrown away his armoured suit. Oh the joys of beekeeping.

The Varroa check of latter years has given way to a new system which not only monitors the mite but also checks the efficiency of the treatment. Thankfully I have not yet found any mites which are resistant to the pyrethroid but none of us can afford to be complacent. My apiary is situated within a mile of an out-of-town "green market". Exotic fruits picked the previous day thousands of miles away in the sunnier countries of the world arrive there daily. An accompanying scorpion or strange spider is not uncommon. One day a small hive beetle will no doubt climb out of a box of grapefruit perhaps to be closely followed by a Tropilaelaps Mite or even a Cape Bee. Oh the joys of beekeeping.

It's one of those lovely gentle summer afternoons that we half remember from childhood years. I've just finished examining some of my hives to satisfy myself that all is well and now have time to relax. Settling down in front of the great stone wall which protects my apiary from the blast of the winter's north wind I can look along the staggered line of hives and watch the back and forth flow of bees. At the entrance to the nearest hive I can recognize the colours of the pollen as the bees carry it in to the colony. The last grains of red from some late flowering chestnut contrasting with the black of a poppy and the green of some meadow sweet and the ever present many variations of yellow and cream. The buzzing of the bees is little more than a murmur and a lark rises in song above the nearby meadow. Oh the joys of beekeeping.

The summer is coming slowly to an end. Most of my bees with fresh supers or section crates have gone to the heather. It's harvest time for me and feeding time for the couple of colonies I still have at home and who will get the wet frames dumped on them to clean up and store remnants of honey for winter. The honey, freed from its comb, runs lazily in a clear golden stream into jars. Oh the joys of beekeeping.

The bees are back from the heather. Not only have they gathered in their winter stores but have given me the profit of some beautifully filled sections. The scent of the heather moors is carried into the kitchen with the section crates and remains with me while I carefully clean and pack each heather filled section. Oh the joys of beekeeping.

It's been a nice year really. The bees, resilient as ever, soon built up into strong colonies. My swarm prevention methods worked; there have been occasions when they didn't. The summer brought the delights of honey flowing into the hives and even the hawthorn produced its nectar on warm summer evenings. We have had good speakers at our association meetings who not only instructed but entertained. We had a couple of nice summer visits to members' apiaries and I had the time to sit and talk to my friends. Yes, it's been a nice year. Oh the joys of beekeeping.

7
The magic of beekeeping (2008)

I really can't remember what happened to 2005-6 & 7 but I didn't enter an essay. Perhaps I didn't like the titles. In 2008 I performed enough magic to get a third.

During a period of over 25 years I have stood at the side of an observation hive at a great many shows describing the activities of honeybees to a wondering and I have to say increasingly understanding public. To them bees have a magic about them because they do so many things which can best be explained as magical. Beekeepers are clearly modern day alchemists whose manipulation of insects conjures golden honey from base flowers. The Arthurian Merlin is alive and well and living in suburbia with a hive of bees in his garden.

To beekeepers the magic of beekeeping is in being privileged to partic-ipate in, observe and understand the activities of what is and will always remain a wild untameable stinging insect.

My bees live in a hive with an entrance about 17 inches long in the front. They know where that hole is in space to the extent that if I move it, say two feet to one side, they can't immediately find it. Navigation that precise has to be some kind of magic. But in August I can move them 15 miles to the heather moors they are aware of the disturbance and re-orien-tate directly to the new location.

My bees have as their head and apparent leader a queen although she is really as much a servant of the colony as any worker. If, for whatever reason, she suddenly vanishes the bees know that she has gone because a magical substance has vanished from the hive in which they live. I know that people call it a "pheromone" but in reality it is a magical potion. Its presence holds a colony of 50,000 bees together. Its absence initiates a course of action by which the bees appoint a leader to the matriarchal throne. The bees can choose a day old larvae which would, if left alone, eventually emerge as a female worker, build a larger cell around it and feed it on concentrated diet of a substance which they exude from a gland in their mouth. The result is that the larvae will emerge as a queen. There is a vast difference between the two creatures. One committed to be infertile and lead a short life with a progressive work pattern that will eventually kill her. The other, much larger, destined to be fertile, head the colony and lay thousands of eggs before separating from it leading a breakaway group, a large swarm, to begin another colony. This change is brought about by an alteration, albeit radical, in diet during the larval stage of its development. Now that has got to be magic.

The box my bees live in contains vertical sheets of wax to which they add wax flakes produced by their own bodies. The vertical sheets of wax, which form the mid rib of the finished storage area, are man made and quite thick. At least they are thick in comparison to the mid rib that the bees, left to their own devices, would make without sacrificing any of the strength.

On each sheet in the darkness of the hive they form, with outstanding ingenuity, hexagonal cells back to back in an overlapping pattern in which the base of one cell shares the base of three others on the reverse side of the mid-rib. This forms a storage area of immense strength and capacity in ratio to the material used. It is also done in such a manner that the sum of the depth of two opposing cells on either side of the mid-rib is greater than the width of the comb. The complex geometry of the design has fascinated mathematicians down the ages. No doubt the packing industry has a name for it but I call it magic.

Enter the magical world of the honeybee with care and a desire for knowledge. Observe the usual rituals of gently applying a little smoke to the entrance pausing a while before removing the roof and any parts above that innermost sanctum the brood chamber. The bees are able to defend their home and family with vicious stings but they also recognize and accept an entry into their domain that does not threaten them so move slowly. One of the snippets of advice which has come down through the ages is to talk slowly to the bees while handling the frames. This has the effect of reducing jerky movements which might bump and annoy the bees inducing stinging but could it not be that the advice originates as repeating some long forgotten spell, an incantation to calm the savage bee.

On the subject of communication with bees is there a beekeeping family who would dare omit the ceremony of telling the bees that their keeper has died. I suspect not. It isn't so many years since I saw a piece of material from a wedding dress tied in a bow and pinned to the front of a hive following a family wedding. There it would hang until it fell of its own accord. Customs built on a foundation of superstition, an attitude which itself is based on a trust in magic.

Instinct, custom, legend, a craft, use any terms you wish in describing the workings of the hive and it's management but you can't deny the fact that keeping bees is just magic.

In 2009 the title was obviously in tune with my own beekeeping and I was delighted to be awarded a First.

8
The art of
coarse beekeeping
(2009)

National Honey Show essay

You must consider carefully before following the path of coarse beekeeping. Its disciples must have the same dedication and attention to fine detail as those who take up any other intricate hobby such as piano smashing.

The first steps of the coarse beekeeper are easy. Your local library will provide you with a copy of one of the many books written by an experienced beekeeper which will illustrate the equipment needed and describe in detail life style of the honey bee. Many experienced beekeepers feel it is incumbent on them to write such a book.

Do remember to renew you possession of the book at the library before fines are imposed, that would never do.

The same library may be able to put you in touch with a local beekeeping association and give you details of their meetings. You should go to a meeting and introduce yourself as being keen to learn about the craft. At this stage a demonstration of enthusiasm works wonders. It might also get you a copy of a beekeeping equipment dealers catalogue. This will save you having to contact one since none, so far as I know, have 0800 telephone numbers. Although allowing yourself to enquire generally about membership and the possibility of free beekeeping classes your enthusiasm should not allow you to actually pay a subscription.

Reading the catalogue together with the beginner's book will immediately convince you that your first pound of honey could be very expensive indeed. However the coarse beekeeper knows that no corner must be left uncut in the search for true perfection.

Your occasional attendance at a meeting, or the hoped for classes, will allow you time to gather up the minimum amount of such essential equipment that can't be substituted by other items. A longish screwdriver and a paint scraper from your toolbox would replace a hive tool. A suitable length of net curtain worn over a broad brimmed hat and tucked well into a jacket could well replace safety equipment such as a veil. A more sophisticated version I have seen is an old fencing mask with further material sewn around it to prevent access by bees. A replacement for a smoker is more difficult unless of course you are a smoker yourself in which case a pipe filled with well rubbed War Horse or a small cigar will suit admirably and yes I have seen it done.

At association meetings always listen for mention of old Harry having passed away or old Jimmy packing up because of his bad back. Here are sources of cheap equipment. Not necessarily good equipment because old beekeepers are noted for putting up with much loved and familiar equipment long after it really should have been changed.

Getting bees is relatively simple. Set out a hive with some used comb in it and wait for a swarm to take up residence. Success largely depends on how far away you are from the nearest beekeeper and could take some time or even fail altogether. A more certain way is to inform local police offices and pest control officers, both of whom are told of swarms having landed in a variety of odd spots, that you are prepared to collect a swarm within a given distance of your home. You should undoubtedly get you some bees that way. Do have a care to check before your journey that they are actually a swarm of bees and not an underground bumble bee nest.

We now look at the management of the bees. It is a fact that the less bees are disturbed by the beekeeper the better they are for it and the more honey you will be able to gather. Disease in bees has become an ever-increasing problem over recent years and must be addressed at all costs. Gone are the days when a coarse beekeeper need only take the roof off a hive twice a year. Once in the Spring to check that the bees flying in and out are actually living there and not robbing and to put some supers on and again in late summer to take off the honey supers. Unless disease is tackled there is little doubt that you will lose your bees. There is of course the short term option of requesting the seasonal bees officer visit you to check your bees. I say "short term" because success in any case depends on what you tell him and I fancy the man will soon whittle out the over-coarse beekeeper who is merely using him so learn quickly from him what you will need to do. The "term" gets very short if you try the old trick of "while you're in there could you mark and or clip the queen for me, add or remove supers" etc?

Otherwise management is mainly concerned with swarm prevention, queen rearing and honey harvesting. Swarm control means far too many visits to and manipulations of the hive and the colony or fiddling about with multi gated boards to suit the true coarse beekeeper. If you allow the bees to swarm in their own time you can save all that work. This also has the effect that you may well be able to collect the resulting swarm from where it rests and put it into another of the late Harry's hives. You will also get a new queen in your existing hive without the bother of all that troublesome queen rearing.

This leaves only the honey harvesting. Although it may be unusual advise for the coarse beekeeper a certain amount of time spent in the preparation will in the long run save both time and money. Buy unwired wax for your honey supers it is cheaper. Cut sheets lengthwise into 4 equal strips and fit one strip at the top of each frame. Only the most profligate beekeeper would use more. The bees will form their own cells along and below these strips. When it comes to harvesting the honey remove the frames, cut carefully along the joint where the bee made cells meet the provided foundation. Cut the oblong block of honey filled comb into sizes to fit cut comb containers or old margarine tubs depending on the destination of the honey. Properly labelled cut comb containers can be sold. That in old margarine tubs can be used to pay any tradesmen prepared to barter his labour for your honey. They are out there, I have had roofs mended and cars repaired.

The coarse beekeeper's preparation of the bees for winter is to go indoors and forget about them until spring. There is no need to mention mouse guards because unless the late Harry had them fitted to the hives when he died the coarse beekeeper is unlikely to own any.

Similarly wasted is the advice not to brush any snow off the hives because it helps to insulate the colony. It would never cross the coarse beekeepers mind to do such a thing.

And so the coarse beekeeper's year ends. If the advise on the unavoidable disease control has been followed the bees should survive the winter. They have after all survived several million of them without the ministrations of "proper beekeepers".

9
Bees and Darwin
(2010)

In 2010 here was a subject to get your teeth into and despite it being quite the opposite of coarse beekeeping it was judged to be my second 1st prize.

If Charles Robert Darwin had not failed in his studies at Edinburgh University he would have followed his father into the medical profession. Instead he concentrated more on his friendship with local fishermen learning so much from their catches that he was able to read a paper on zoological observations to the university's natural history group, the Plinian Society, in 1826 when he was only 17ys old.(1)

If, at Cambridge, he had not been distracted from the classics and Paley's Natural Theology he would have entered the church. Instead he joined the hunting and shooting set and collected butterflies.(1) Most importantly he befriended Professor Henslow the botanist.

It was this friendship which ultimately led to him being invited to join HMS Beagle as a naturalist on its five year voyage from 1831 to 1836.

For the length of the voyage Darwin does not seem to have done any work on bees or at least none is recorded in "The voyage of the Beagle". Once settled in Down House, Kent, in 1842 he engaged in the common Victorian hobby of beekeeping and began to study them.

His main interest was in the complex construction of the cells. This was a matter which, although bees had been observed for centuries, was not fully understood. At one point he followed the idea put forward by his old associate George Waterhouse(2) that the bees formed a cylindrical cell from which hexagons were formed by the external pressure of surrounding cells. He knew this to be the case with the Mexican stingless bee Melipona Domestica. Darwin hoped to be able to show evolutionary steps from the simple spherical cells of the humblebee through M. Domestica to the complex cell of the honey bee. He spent so many years studying this problem that one commentator was led to suggest that Darwin had delayed his publication of The origin of the species until 1859 to give himslf more time to prove his point. (3)

Darwin's work on honey bees spanned the period during which Langstroth realised the importance of the "bee space" which led to the introduction of the movable frame hive and the creation in 1857 of pre-formed foundation. T. W. Woodbury who introduced moveable frames into Britain in 1862 sent Darwin a sample of this foundation. In a letter to thank Woodbury Darwin wrote that it "interested me much" (4)

In studying cell building he introduced pieces of wax including some dyed with vermillion and gave lengthy and detailed descriptions of his experiments and observation in the Origin of the species. (5)

He was considerably helped in 1858 by the installation of an observtion hive with the assistance of the Rev John Innes who stocked it for him. Surely a true example of christian charity since, on the face of it, Darwin's work was aimed at proving the theory of Evolutionism against the theologian theory of Creationism and, en-route, largely doing away with God.

The observation hive enabled the scientist to study comb building at close quarters allowing him to complete his great work, "On the origin of species by means of natural selection, or the preservation of favoured races in the struggle for life". published in 1859.

In the end he was unable to find the evolutionary steps he sought and in his Origin of the species he wrote, "The subject of instinct might have been worked into the previous chapters; but I have thought that it would be more convenient to treat the subject sepa-rately, especially as so wonderful an instinct as that of the hive-bee making its cells will probably have occurred to many readers, as a difficulty sufficient to overthrow my whole theory. I must premise, that I have nothing to do with the origin of the primary mental pow-ers, any more than I have with that of life itself. We are concerned only with the diversities of instinct and of the other mental qualities of animals within the same class." (6)

On this matter I fancy he came as close as possible to describing it as an instance of one use of the word teleology "the doctrine that there is evidence of purpose or design in the universe, and especially that this provides proof of the existence of a Designer"

Darwin studied humble-bees as they were known at the time by observing them at great length in and around the garden at Down House and walking the local hedgerows, He was well aware of the importance of both honey bees and humble-bees in the cross pollina-tion of plants and wrote extensively on he subject. (7)

With the help of his children he particularly studied the move-ment of male humble bees around his garden. He found that they tended to follow set paths which he recorded in his field notes. (8) He also found that they had chosen places along their route at which they paused and "buzzed" before movong on. On 17th September 1857 he wrote "Saw several on the same thorn … … I saw what I fancied was a female come and find with difficulty the tree and it rested on twigs and seemed to sting them" At this point he made a marginal note "Do not females find males by their buzzing places?" Was he describing what in the honey bee world is known as a drone assembly area? (9)

In the matter of pollination one of his most important observations was bringing attention to the fact that only humblebees could pollinate red clover which had not previously been known. He wrote "The tubes of the corollas of the common red and incarnate clovers (Trifolium pratense and incarnatum) do not on a hasty glance appear to differ in length; yet the hive-bee can easily suck the nectar out of the incarnate clover, but not out of the common red clover, which is visited by humble-bees alone; so that whole fields of the red clover offer in vain an abundant supply of precious nectar to the hive-bee."(10) This observation led directly to humble-bees being taken to New Zealand to pollinate red clover when it was found that there was no indigenous insect with a tongue long enough to reach into the flower. The first attempt was made in 1875 but it was ten years latter before a successful transfer was made. (11)

Darwin was also aware that the population of humble bees in an area was largely reliant on it's nest not being eaten by mice. On this he wrote "I have, also, reason to believe that humble-bees are indispensable to the fertilisation of the heartsease (Viola tricolor), for other bees do not visit this flower. From experiments which I have tried, I have found that the visits of bees, if not indispensable, are at least highly beneficial to the fertilisation of our clovers; but humble-bees alone visit the common red clover (Trifolium pratense), as other bees cannot reach the nectar. Hence I have very little doubt, that if the whole genus of humble-bees became become very rare, or extinct or very rare in England, the heartsease and red clover would wholly disappear. The number of humble-bees in any district depends in a great degree on the number of field-mice, which destroy their combs and nests; and Mr. H. Newman, who has long attended to the habits of humble-bees, believes that "more than two thirds of them are thus destroyed all over England." Now the number of mice is largely dependent, as every one knows, on the number of cats; and Mr. Newman says, "Near villages and small towns I have found the nests of humble-bees more numerous than elsewhere, which I attribute to the number of cats that destroy the mice." Hence it is quite credible that the presence of a feline animal in large numbers in a district might determine, through the intervention first of mice and then of bees, the frequency of certain flowers in that district!"(12)

There is one story which is inseperable from any commentary about Darwin and humble-bees The local farmers noticed a steady decrease in their crops so decided to take their problem to the great Darwin now settled in at least semi-retirement. Darwin agreed to look at the situation. Together with a group of farmers he walked some of the hedgerows before declaring "Get more cats" then turning to walk away. The farmers did not understand the instruction and called after him to explain. "Your crops are fertilized by humble-bees. Their nests are being eaten out by mice. "Get more cats which will eat the mice and allow a larger population of bees" and continued on his way.

On a lighter side Thomas Henry Huxley another eminent English biologist, and others went on to expand this idea, pointing out that the success of the British Empire really depended on "Old Maids". Why? ... because soldiers eat roast beef, the beef cattle eat red clover, red clover is pollinated by bumble bees which, in a round about way are protected by cats. The cats eat the mice that prey on the honey and Old Maids keep cats, therefore the continuation of the British

Empire was really dependant on cat loving elderly ladies.

It's a nice thought and nice enough for A. D. Hope to write a poem about it. In "Clover Honey" from "A Late Picking: Poems 1965-1975", he includes the following verse:

So when we find - what does the Bible say?
A land flowing with milk and honey, we do
Not doubt, we naturalists, that there we may
Expect to find old maids a-plenty too.

Darwin's study of bees may not be the first thing about his work that springs into the mind of the public or for that matter the minds of beekeepers. His knowledge and studies of matters concerning natural history have without question made him a world renowned figure. One has to wonder however if, had he entered medicine or the church, his eminence in either field would have been such that he might enter anyone's mind at all.

References

Nora, Lady Barlow. Charles Darwin's granddaughter submitting a biography in Chambers Encyclopaedia.

George Waterhouse (1810 – 1888) a British naturalist who Darwin had engaged to assist in the study of the mammals and beetles he had collected on the voyage of the Beagle.

Robert Richards, "Why Darwin delayed, or interesting problems and models in the history of science" Journal of the History of the Behavioural Sciences Vol. 19. Jan 1983.

Letter in the hands of the beekeeping museum of the Bee Research Association (Bee World Vol 40 No 12 Dec 1959.

Darwin, C. R. 1859. On the origin of species by means of natural selection, or the preservation of favoured races in the struggle for life. London: John Murray. Chapter 7. Page 228

Darwin C.R. 1859 Origin of the species … … Chap 7 page 207

Darwin C. R.. The effects of cross and self pollnation in the vegetable kingdom (1876)

R.B. Freeman. Charles Darwin on the routes of the male humble bees. Bulletin of the British Museum (Natral History) Historical Series. Vol 3. No 6. 1968

B. A. Cooper. Have you heard a drone assembly. BIBBA 1997

Darwin, C. R. 1859. Origin of species … … Chap 4 page 94

Bumblebees to New Zealand By Dr David Sheppard, Natural England Invertebrate Ecologist, 21st May 2009 www.bumblebee-conservation.org.uk/Bumblebees_to_New_Zealand…

Darwin, C. R. 1859. Origin of species … … Chap 3 page 74. quoting from Newman H.W. 1851. Habits of the Bombinatrices. Proc. ent. Soc. Lond., meeting on June 2nd.

10
Disastrous
beekeeping
purchases
(2011)

I did have a problem in 2011 because I had never made a disastrous purchase unless of course you count buying my first ever hive which got me into this mess in the first place. However I knew someone who had made a purchase which proved somewhat disastrous so I swopped places with him for the occasion but only came second.

Once upon a time when I was much less knowledgeable in the ways of bees I made a wonderful money saving deal which cost me dearly.

It was probably only year two in my beekeeping life but one of the things I had mastered was the collecting of swarms as a means of acquiring free stocks of bees. The difficulty was that the number of swarms I collected used up my available equipment faster than I could afford to replace it.

The chaps in my pub kept asking, "How are the bees?" Not that I believe they would have understood the answer it's just that they didn't quite know how to take a man who kept little stinging insects as a hobby. So on Monday night I decided to answer the question with my problem of the lack of brood boxes. People really are interested in bees and my problem was relieved by one of the lads, John the joiner, offering to "knock a couple up" if I showed him what they where like.

The next day he called and I showed him a national brood box which he measured and sketched. I took the roof off to show him the internal arrangements. The bees were quiet enough and although he did look into the hive as the buzz rose in pitch he decided that that was enough and he could see how it all went together.

On Friday evening I saw John in the bar and he told me that he would bring the brood boxes over in the morning. I bought him a pint and looked forward to resuming swarm collecting.

During the week I had busied myself making and waxing up frames so that I was ready when John delivered two brood boxes. They fitted exactly onto floors and a roof sat neatly on them. I paid him willingly, although a little more than I had expected. Still it was less than the major dealers would have charged, there was no carriage and it is nice to support local tradesmen, isn't it. I filled the first box with ten frames, I actually thought I would have got eleven frames in. And ten in the second box, must be right.

A few days later when both were filled with swarms my thoughts turned to the fact that the harvest would pay for the brood boxes. That summer those bees joined the rest of the apiary by apparently going on strike and producing little or no excess. That was also the year I learned beekeepers had a saying, "There's always next year"

Next year brought a visit by the seasonal bees officer checking for foul broods. Varroa had not, by then, graced our shores. He worked steadily through the hives until he came to the new ones. Off came the roof, the super, the queen excluder and, deftly using his J hive tool, all the top bars on the brood frames.

"You've nailed these up wrong. The nails go through sideways not down from the top"

"Oh dear" I said meekly.

Again plying his hive tool he tried to loosen a frame by levering it sideways. Nothing happened. Peering into the hive he said "Oh I see what's happened. Better come and have a look at this."

By the light of his torch I looked down at the ends of the frames as he directed. I had learned by now about the "bee space" and where I was looking there wasn't any.

Propolis is the bee's glue of choice. Collected from trees by bees towards the end of their life it is taken to the hive and immediately used to block small gaps. I was looking at what used to be small gaps. Lots of them.

John the joiner had taken the external measurement of eighteen and one eighth inches, noted the general construction and made two hives exactly to those specifications. The problem was that he hadn't noted the thickness of the wood used. Instead of the walls being three quarters of an inch thick, they were one inch thick. The knock-on effect was to rob the internal space of half an inch. The brood frames which should have hung leaving a comfortable three eighths of an inch at either end now hung leaving only one eighth of an inch at each end. One eighth of an inch is a space which, being of no use to a bee, they fill with propolis. And that is exactly what the bees had done to both sides of both ends of ten frames in both hives. The same reasoning applied to the width which is why I could only get 10 frames into the box.

"Well if you can ever get them out I'll come and examine them. Good luck."

I quickly rang a dealer for 2 new brood boxes plus frames plus wax. That hurt.

Eventually the bees were tempted out into new boxes and I managed to remove some of the frames intact from the wrong sized boxes. I rose above the disaster and learned from it such as how to nail frames up, that was a silly mistake. John the joiner has made me more equipment after I gave him the plans and has made himself a couple of hives too. Since then he is greeted with the familiar "Now then John how are the bees?"

Isn't there a saying about clouds and silver linings.

11
A funny thing happened at the apiary (2012)

I do like a title which includes the word "funny". It save me having to think of how to keep it light hearted. But it was only funny enough to come second.

A funny thing happened at the apiary today. Well you need a laugh now and then. It's not all fun being a queen especially one with a small q. Outliving all the kids is a bit of a downer for starters. They seem to be dropping dead left right and centre these days. Then we've got this great creature with a smoking dragon thing that comes poking around the house usually leaving everything in a real mess that the girls have to clear up. So you see we need to make our own amusement generally by trying to mess the creature about or getting the odd sting at him.

I remember a couple of years ago when I wanted to move house. The girls built a few cots for me to lay eggs in. Well you've got to leave the place with someone to take charge haven't you. I like to call them the princesses royal. No sooner had I finished than the creature came bursting into the house and knocked them all down. The girls did no more than start building new cots. I laid the eggs and told the front door keepers to watch out for the thing returning. The eggs hatched out and the little babies were doing nicely when back he came and did the same thing. What a waste of effort. Anyway we had a good look around the house and found a corner of one of the bedrooms where the creature had broken the wall and the girls had to repair it. I knew I should never have set those young'uns on to do the job. They made a real mess of it, wasn't straight at all, in fact they left it with a big fold in the wall. Just big enough to build a cot out of sight. So we did the whole cot building thing again, laid the eggs and waited. Sure enough, just before the cots wanted their covers on them, back he came and just squashed them. Ah but he didn't see the one in the fold. Well done the sisters. Two days later it got it's cover on and we were away.

About half of the girls and some of the lads took off with me and landed on a tree just opposite the house. A couple of the girls who had gone for a scout around came back and told me that there was an empty house just a couple of doors away and they didn't see the point in going too far so we moved in there. Mind you we soon found that we hadn't got rid of the creature.

Now where was I? Oh yes, today. The creature came back a couple of days ago poking around the house filling the place with smoke so you couldn't see across the room. And he's gone and taken the pantry. Just got it nicely filled up, the lids put on and its gone. The girls had been back and forth to the yellow fields for days and days. There's a lot of work in getting a pantry filled. Gone, without a "by your leave" or word of thanks,

The girls now tell me that he's damaged the ceiling so two of the little bars now have a big hole between them. Don't know what we are going to do about that. It's big enough for the boys to get through not that that matters too much. It's not as though they were going to go through and pinch stuff you've got to push food into them.

Today the creature came back and put the pantry back on top of the house. Mind you it's empty and a real sticky mess. The girls are busy clearing it up, consolidating our assets you might say. I went to have a look because I can now get through the gap the creature made in the ceiling. It's quite nice there so we've come up with a cunning plan. Lets really mess the creature about. If the girls can get this food out of the way I reckon I can have the whole box filled with eggs in about 5 days. Then lets see how keen the creature is to pinch it.

12
My definition
of a successful beekeeper
(2013)

To be honest I struggled with this one. Just how do you define a successful beekeeper? I was delighted to get another second prize.

Measuring success may have different meanings to different people. Who harvests the most honey? Who can most successfully breed good queens? Who has the calmest bees that still gather in loads of honey? All these are of course indications of success but I think we must look deeper and wider than that to define my successful beekeeper. Beekeeping is a craft of many parts and the truly successful beekeeper must have as many layers of expertise.

Knowledge of the bee. Without which you can be little more than a "leave 'em alone" beekeeper. I don't mean a need to pass all the

BBKA examinations, but a beekeeper of any quality must have a sound understanding of the bee's life style. An understanding of what triggers which instinctive reactions. Certainly the greater the knowledge the better the chance of travelling further along the path of success.

The bee colony works between parameters of ability beyond which it either will not or cannot operate. The beekeeper needs to understand these to ensure that, because of his manipulations of the colony and the hive he does not force the colony into an impossible situation.

By now the aspiring successful beekeepers will have rid themselves of any colony showing undesirable traits and settled for a quiet gentle non-following "successful" bee.

Patience. Very experienced beekeepers react to given situations almost instinctively. They arrived at this position by being ready to stop and think instead of dashing into something and risk getting it wrong. In any event you can't successfully hurry bees into doing anything. The nature of the beast just doesn't allow it.

Observation. I know beekeepers who scornfully accuse others of trying to understand what is going on inside a hive by standing looking at the entrance. They have obviously never "observed" the entrance and seen just how much information can be drawn from it. In any case it is a good spot to start from. At least finding chalk brood inside the hive shouldn't come as a great surprise you should have seen it at the entrance. Observing, as opposed to simply looking, has to continue throughout the examination of the hive. Keen observation and the ability to interpret what you see improve a beekeepers ability to be successful.

Confidence. Confidence in one's ability gained from practise to open a hive of 50,000 stinging little insects. Then to handle them gently and caringly in such a way that you know they will not respond by attacking you en-masse. The confidence that, if something does go wrong and the bees do get too agitated, you can deal with it calmly and efficiently.

Staying power. Anyone who has been in this game for more than 4 years is aware of the set backs that are to be endured, overcome and from which we need to bounce back. This really is a hobby in which we can always look forward to next year. Dare I say sometimes year after year? I can look back on fairly recent years when honey classes in local agricultural shows were cancelled because of a lack of honey. Nevertheless beekeepers stick with it and next year it sometimes comes right. I remember 1987 when, across the country, about 50% of bees were lost. In my area only some very old beekeepers took the decision to bow out gracefully. It took time but beekeepers and beekeeping did bounce, perhaps too energetic a word, back

Luck. We could all use a bit of good luck throughout the beekeeping year even if it is just in the weather conditions that are thrown at us. We have all heard the relief in a beekeeper's voice when he says that his bees "have survived the winter". I was at a meeting in September 2012 where I heard the same relief when someone said that his bees had "survived the summer". I would settle for a lack of bits of bad luck.

Some beekeepers consider themselves "lucky" enough to live above the oil seed rape level. Their bees can gather a crop of light delicate dandelion honey. Some of them have bees located where they can gather a heather crop by the simple expediency of turning right towards the moors instead of left towards the meadows. All very well but does it add up to true "success".

Winning prizes at honey shows. A wall covered with impressive prize cards from national or local honey shows is visible proof of success in beekeeping. It's presence could very well count towards being an all-round successful beekeeper but I would argue that it's absence should not go against an applicant for the position. I know very successful beekeepers who are just not interested in the competition of the honey show. Local honey shows in particular are our shop window to the public and for that reason I like to see them supported but I am happy to respect the opinion of those beekeepers who do not wish to join in.

To sum up, given a bit of luck, any keen, knowledgeable bee-keeper who has the confidence to gather swarms, and manage his bees with care and understanding could well qualify to join my list of successful beekeepers.

Having acquired the appellation of a successful beekeeper I would expect them to round it off by putting something back into beekeep-ing by teaching or mentoring the less experienced and guiding them towards success.

13
Overheard in the hive:
"Our Queen has gone missing,
now what shall we do?"
(2014)

Errr what on earth was I going to do with this. Well it had to be Alice in wonderland and the rabbit didn't it. And good enough to get another second, jolly good.

It was a bit of an Alice in Wonderland moment really. I was sitting on a stool in the warm sun and leaning quite comfortably against a stonewall, watching the bees coming and going. I suppose I should check them for queen cells but I was just beginning to wonder if it was actually worth disturbing them when I heard some voices. Bit of a babble really some way off or was it? No it was coming from the beehive. That's silly I thought but moved my ear nearer to the hive. The babble went on. Clearer and more anxious than the usual buzz. Then a voice rose above the rest, "I know what's wrong. Our queen has gone missing, now what shall we do?"

"Yes, you know what, I noticed a difference when I came back from getting water when the sun was high." Said another voice. "The place smelled different. The scent was just less intense in fact now it's almost all gone"

The voices were stilled. Just for moment as though a thought was going round the colony. Then came the cry of "Look for her, look for her". It was taken up and repeated in all parts of the hive. Bees scurried out of the entrance, not to fly away but to crawl all over the outside of the hive. The searchers spread out, across, over and even under the hive. Eventually the searchers returned to the entrance and vanished back into the hive.

"She's not there" was echoed through the hive.

"Well, come on. Anyone got any ideas? Someone called out.

"I hope she hasn't been tempted to go out for a fly with one of those drones, you know what they're like it's the only thing they live for."

"She can't fly she's had the snip." Said someone rather dismissively.

"Wonder if she tried and fell into the long grass. She'd get so far out even after the snip."

"Has anyone looked?"

"Are you joking we're not going poking about in that stuff. Chances are we'd never get out ourselves. Anyway if one of us found her we couldn't carry her back."

"Well we are going to have to do something."

"What would you know you've just been born"

"I know but I seem to have some feelings about it all."

"Oh we all have them but without the queen they don't always get put together right"

"What about a committee?"

"Whose she been talking to? Stupid bee, what does she think we are, a democracy"

"Just a minute has anyone seen any of those big cells hanging down anywhere?"

There was a chorus of "no"s from various places. Then someone said, "But there are a few of those silly little cots knocking about that she keeps getting us build and then doesn't use them"

"We had better make sure she hasn't laid in any of them before she went"

"Yes" said another, "And look in all the bits where the repair jobs have gone wrong. I always said it was silly getting the kids to do the repairs. You could hide a wasps nest in some of messes they made. Off you go."

"You do realise that if she hasn't laid in the little cots we are not shifting eggs about to put into them. It's just not done"

"Oh we all know about that old fairytale. We'll soon empty some cells round a decent looking egg and build a proper cell round it."

"Better do two or three"

"I prefer the idea of doing it to a larvae"

"But if you pick the wrong larvae there isn't time to properly feed them into a queen"

"How do you know?"

"I don't know how. I just do so there"

"Well as soon as that lot get back from checking the little cots and looking in the corners we'll have to get a move on"

"What's the rush?"

"Because you haven't got all the time in the hive to lie about doing nothing."

"No 'cos the eggs will go on hatching and we don't know when she laid the last ones"

There was a moments increased buzz and one of them said, "Right we've been round the little cots and there's no eggs in any of them."

"And we've been round all the dodgy repair jobs and there's definitely no big cells or even cots in them"

"So shall we get started then?"

"Just a minute what happens if we don't bother?"

"Don't even think about it 'cos none of us really know"

" I only know that we've got to do something"

"Yes I feel that as well. Come on girls lets choose some spots to create a queen"

This seemed to raise a real buzz for a moment which soon fell to a working calmness.

I suppose it was about this time that I had the crazy thought of looking around in case there was a bottle nearby marked "Drink me". But then, if there was, I wondered if I would be able to resist the temptation of drinking it and if it did have the effect of making me so small that I would be able to squeeze through the hive entrance what trouble would I be in? I suspect a small humanoid figure wandering past the guards would attract unwanted attention. They might even sting me to death and then, treating me as they might a dead mouse, isolate my body under a sanitizing covering of propolis. Imagine the stir that would cause next year when the new owner of the hive did his Spring-cleaning.

I shuddered not with fear but the realisation that while I had been concentrating on eaves dropping the bees the weather had changed and a big black cloud was above me. Well that finished any ideas of looking into the hive today.

But tomorrow I really must. I wonder what I'll find?

14
The importance
of Queens
(2015)

Had to rather get my technical head on for this one. It seemed to demand it and it got me my third first prize card.

The importance of this, the largest member of a bee colony, has been acknowledged though not necessarily understood since the earliest times that bees were observed and those observations recorded. We must forgive the 1st century Pliny calling it a king1. A reasonable conclusion to draw at a time that even described the drone as a female. It retained that incorrect rank until the 17th c. when Jan Swammerdam using the newly invented microscope identified it as female.

Those beginning beekeeping soon learn the importance of a "good" queen. How often do we hear in answer to a remark about a hive the words "change the queen"? "I'm finding a lot of chalk brood", "Change the queen"; "I've got one showing signs of Acarine" "Change the queen"; "That hive in the corner really is angry", "Change the queen". And so it goes on. It is the panacea of the apiary.

She is the mother of the colony though not as her title would suggest its ruler. She is subject to the demands and requirements of the colony. The queen alone is the layer of eggs. She is attended in her labours by a retinue of worker bees who are her carers in that they feed and groom her but they also shepherd her to lay according to the demand of instinct within the colony. It is that which requires her to lay unfertilised eggs that will hatch out as male drones or fertilised eggs which will hatch out to be female workers. In times of plenty she will be encouraged to lay more while in times of dearth her laying will be discouraged. In Spring she will be encouraged to lay an adequate number of unfertilised eggs to produce drones and in late summer none.

It is in her egg laying that she brings various attributes to the hive. Not all of them welcome.

The queen is the product of and carries the hereditary qualities of her mother and "one" of several drones which mated with her mother. To her "sons" she passes on qualities of her own which include those of his grandparents but are not affected by any of the drones which mated with her. Drones don't have fathers only grandfathers it follows of course that they have no sons only grandsons 2. To each of her daughters she passes on her qualities as determined by her parents and one of the drones which mated with her. Each of her daughters may not be full sisters but because of the complexity of multiple mating will be half sisters. This doesn't actually drive a cart-horse through Mendalism it only makes the mathematics more difficult.

It also means that getting a "good" queen is essentially a matter of either "making" one by controlled breeding through artificial insemination or flooding your area, preferably an isolated one, with bees and drones of your choosing and hoping for a good mating with your virgin queen. Artificial insemination may well guarantee the quality of the offspring but in Bro. Adam's experience it shortened the useful life of the queen3. It also requires the use of some extremely expensive equipment or the skill of a model engineer to make it and the steady hands of a surgeon to operate it. Increasing the number of drones in your locality will be done naturally by the colonies at the appropriate time but the beekeeper can always give them an encouraging hand by inserting some drone comb into a chosen hive.

The attributes which are mainly sought after include the following;4

In the new queen a fecundity which, together with the industry of the workers, will provide the strength of a colony to gather the best honey crops. But it must be of a strength that does not fail mid-season. The work that a colony under a newly mated queen displays in the first weeks is quite remarkable but how soon and to what level does it reduce? In early September enough young bees must be reared to carry the colony though the winter.

A resistance to disease in the colony. As I have already pointed out there are queens which raise bees with a greater resistance to Acarine than others. This will have become clear to the beekeeper among his colonies and will surely have been taken into account in choosing breeder colonies. Breeding the same resistance against nosema is more complex and is linked to the vitality of the colony and it's tolerance to the mite.

In carefully choosing colonies from which to breed bear in mind those which will have been thrifty enough to have gathered enough stores and be hardy enough to survive one of our colder winters.

A disinclination to swarm. When I find colonies which seem to swarm at the drop of a hat I really wonder if we are not still suffering from the old beekeeping practice of killing off those colonies which did not swarm and keeping those which did. All methods of swarm prevention are time consuming and better avoided if possible by raising bees from those colonies which do not rush to fill the brood box with queen cells at the first sign of Spring.

The comfort of the beekeeper and, depending on the location of the apiary, the safety of the public should not be forgotten in the choosing your better bred bees. We should always be looking for a colony of good temper and quiet disposition. I imagine that long before embarking on the road to queen rearing the beekeeper will have disposed of those colonies which lift violently from the comb, bounce off your veil and follow you all the way back to the car. I have heard it argued that bad temper in bees is related to the colonies honey producing performance, certainly it describes a colony which will deter robbing by man or beast. To support that argument I did know a beekeeper who, years ago, bred a colony so calm that it could be demonstrated at schools without fear of the audience being even troubled by them. The down side was that he had to feed them all summer. This may be carrying things a bit too far and in any case I know I am not alone in having colonies which will store loads of honey but will largely ignore my intrusion into the hive.

There is no denying that keeping bees which don't rise from the frames and largely ignore you as you go through the colony is an absolute pleasure. The opposite with angry following bees should not be tolerated. The answer, "change the queen" which is more or less where we came in.

References

1 The lore of the Honey Bee. Tickner Edwardes chap 1

2 Breeding the honey bee. Bro. Adam. English edition 1987. p32.)

3 I really don't know if he recorded this opinion anywhere. I haven't come across it but he certainly expressed it in conversation with his beekeeping friends in NE England.

4 op cit Part 2. Readers should also refer to this book for more detailed descriptions and minor traits which might be transferred in bee breeding.

15
Why don't we just let the bees swarm; it is the natural thing for them to do?

In the summer of 2011 the Sharon Blake the editor of the BBKA News rang and asked if I would do a series of articles for her. I had just had two 1st prizes at the National. When I found out that what she wanted was the monthly instructional article on "what to do in the apiary this month" I graciously declined as being nowhere near qualified to lecture the country's beekeepers. I did say that I could usually manage to write something to a given title. Not long after that she sent me the following title and asked for a somewhat daunting 1700 words. She must have liked it because she did publish it.

Why not indeed. The very first reason I ever heard was the somewhat glib "because we are beekeepers not bee-let-em-all-fly-away-ers"

Swarming is certainly an essential aspect of the life of honey bees. A phenomenon noted and written about by Homer, expanded on by Pliny and included in every bee book written ever since. Not everyone got it right and the texts are peppered with accounts of myth and rumour. Does anyone these days beat a large pan with a heavy spoon to claim ownership of a swarm they are pursuing? Is there some dark deep Devon valley where the first cast is still called a colt and the next a filly?

There are three occasions when the colony will replace its queen without human intervention. 1, In an emergency; 2, By supersedure; 3, By swarming.

1. In the event of the sudden and unexpected loss of the queen the colony will quickly move to build one or more queen cells queen to rear a successor.

2. I have heard supersedure described by beekeepers as a much more desirable trait in their bees. In fact all the bees have done is replace an old and failing or injured queen. But in any event what is important is that the "colony", which has been described as immortal, survives. If, in its evolution, the honey bee had gone along the path of only superseding queens then at some point the creature would have failed. Lost colonies would not have been replaced and the honey bee would have dwindled away into extinction.

It has even been suggested, I suspect due to a bit of anthropomorphisim, that the eggs from an old and failing queen would not provide a good basis for a strong new queen. The effect of age on humans is not reflected in bees.

Whilst bees have obviously always swarmed I suppose we might be allowed to wonder if old beekeeping practices did anything to change the bee which we have now. Did the many years of keeping bees in skeps and, in order to harvest the honey, the practise of killing off those colonies that had not swarmed but might have superseded, leave us with a "swarmier" bee?

3. And now we come to the nub of the matter. Swarming. Whole books and complete chapters in other books are dedicated to the subject. They explain in minute detail a wide range of procedures to follow in order to delay or eradicate the event.

For those who find the need to practise some form of swarm prevention there are many methods, and variations on methods, available. If the prevention of an instinctive process is to be carried out successfully then the most sensible way seems to me to be one which keeps the gathering force together and convinces the bees that they have swarmed.

Swarm control

Probably the first and apparently simplest method, though possibly the worst, a beginner comes across is the destruction of queen cells. It is the worst for a number of reasons. It is too easy to miss a queen cell tucked away in a corner. It demands regular visits, usually 7 days apart, to repeat the exercise. This repeated destruction causes stress to the bees and in any case the mathematics are wrong. Eventually the frustrated bees will choose an older larvae to develop into a queen and swarm ahead of the next visit.

Some other methods, include the use of designer boards which either divide the colony or interrupt the gathering of nectar and honey production. Many have the names of the authors or designers attached to them, a veritable who's who of the Edwardian beekeeping world. The timing of some manipulations can have a knock-on effect. Eggs laid between, generally, the 20th June and 10th July hatch out to become the work force on the heather moors. Those dates were given by Bro Adam and I suppose relate to Dartmoor. In the North East of England one author extended the later date to 20th July thus giving the last bee to emerge two weeks of foraging.

To that list I am going to add the following method practised by a Durham beekeeper who keeps two of his hives on scales and can verify the continued income of nectar and production of honey through-out the process. This at a time when major disturbance and splitting of colonies would normally reduce it.

Swarm control without weight loss

To carry out this method the beekeeper will need a brood box and a specially modified crown board along with 10 frames of foundation or drawn comb.

The crown board has a 4 inch diameter hole cut in the centre covered with fly mesh. An entrance is made at one side of the board. The advantage of the 4 inch diam. hole in the crown board is to allow heat from the parent colony to rise up and aid rapid build up.

When signs of swarming are found in the parent colony the queen must be found and temporarily caged. Find a suitable frame with one or two unsealed queen cells, two frames of sealed brood and one frame of stores. These, all with the adhering bees, are placed in the new brood box ready to be put above the modified crown board.

The frames taken from the brood box are replaced by four frames of foundation. The old queen is released back into the parent colony, a queen excluder placed on top of the brood box and supers added.

The modified crown board is placed on top of the supers with the entrance at 90° to that of the parent colony. The brood box with the queen cells, sealed brood and stores, to which 6 frames of foundation or drawn comb has been added, is placed on the crown board. A quilt and roof are added and the whole left for three weeks.

Heat from the parent colony will help the new queen to build up rapidly. The heat together with the rising scent results in a smaller loss of bees returning to the parent colony than might be expected. The mesh will eventually be filled with propolis but not until after it has served its purpose. After 5 weeks the old queen can be removed and both brood boxes can be united resulting in a strong colony which can be taken to the heather.

This swarm control method has been carried out on a platform scale thus measuring the daily weight gain or loss when the weather is poor, winter and summer

The arrival of Varroa heralded the end of those halcyon days of leave-alone beekeeping, a time when it was only necessary to take the roof off a hive twice a year. In the spring to see if the bees flying in and out were living there or just robbing it and again in the summer to take off some honey. Those beekeepers got lots of honey because in the early summer their hives were not being continuously interfered with in the name of swarm prevention. Yes their bees swarmed but the use of prepared and suitably placed bait hives probably meant they collected most of their swarms so losses could be replaced.

Changes in varroa control and hive "cleaning methods" have become easier. As the name implies, the shook swarm method controls the varroa, and used at the right time can satisfy the colony's instinctive urge to swarm but do read the instructions. The use of oxalic acid is less disturbing than the Bayvarol strips of the 90's.

But the question I am set is - should we be using them at all. Perhaps the answer is to be found by looking at the location of the bees themselves.

How pleasant to sit on a summer's day with the humming of the bees and watch a clouded swarm pass over …… on its way to terrify the wits out of some unsuspecting soul. We owe it to the public and indeed to ourselves and other beekeepers to be responsible beekeepers. Be honest a loose swarm of bees in an urban area is a nuisance. It brings mayhem to a street and can bring the centre of a small town to a halt and don't the press love it. Yes the urban beekeeper should make every effort to stop swarming. At the same time beekeepers should be prepared to "be the hero" in the eyes of an admiring public and collect swarms from positions where it is safe and practical to do so. I think we all accept that swarms do get into locations where the only option is to call in the pest control officer. I have known pest control officers who became beekeepers having been called out to destroy wasps which turn out to be honey bees and, rather than kill them, have boxed them up taken them home and acquired a new hobby.

There was a time, albeit twenty years ago, when swarms were collected without fear. Any Varroa were on the bees and could easily be dealt with. The fear came from the swarm that was not collected but vanished into the depths of a hollow tree from where varroa could re-infest hives. Modern thinking is rightly more cautious with the findings that swarms can carry AFB and EFB. Advice now is that collected swarms should be quarantined until their health is verified. A swarm emerging unchecked from the truly rural out apiary may not cause the same amount of public disturbance but the beekeeper's work is not over. Somewhere there is a fairy tale world in which the bees either only make one queen cell or, if more are used, the first queen to emerge will rush around killing all her unhatched sisters.

There is a beekeeping choice. Option one. The beekeeper can either put work into controlling swarming. Or, option two, having allowed a swarm, immediately put the effort into carefully searching the swarmed hive and reducing excess queen cells making sure that the colony does not "cast" itself into extinction.

You can please yourselves. I vote for option one.